ITALY

José Gelabert-Navia

University of Miami

OSCAR RIERA OJEDA
PUBLISHERS

Preface

In 1997 Kathy Eldon, the mother of photojournalist Dan Eldon published a book called ***The Journey Is the Destination.*** The book contained 17 journals, replicated collages of writings, drawings, and photographs. Eldon was stoned to death in Somalia in July 1993 by a mob reacting to the United Nations bombing raid on the suspected headquarters of General Mohammed Farah Aidid. He was only twenty-two.

I remember when the book came out. I was taken by the depth of his understanding of the places where he lived and worked in, notably East Africa. I was also taken by what he said was his mission:

"To explore the unknown and the familiar, distant and near, and to record, in detail with the eyes of a child, and beauty of the flesh or otherwise, horror, irony, traces of utopia or Hell."

My own notebooks and sketches do not record horror or hell.... But they do aspire to remember in detail and with the eyes of a child places visited and journeys made. Most of these drawings are the result of what became annual pilgrimages with my students and with my family. For me, the cities always appeared first through literature and only then through its architecture.

William Carlos Williams, who would spend most of his life in Patterson, New Jersey, would write in ***January Morning:***

I have discovered that most of
the beauties of travel are due to
the strange hours we keep to see them:

the domes of the Church of
the Paulist Fathers in Weekhawken
against a smoky dawn—the heart stirred—
are beautiful as Saint Peters
approached after years of anticipation.

And so, it has been for me. These drawings have captured an instant, an image, to be collected, saved, and retrieved years later to have the memory glow again.

Jose A. Gelabert-Navia
Josegelabert-navia.com

ITALY

*"To bless this region, its vendages, and those
Who call it home; though one cannot always
Remember exactly why one has been happy,
There is no forgetting that one was."*

W.H. Auden, *Goodbye to the Mezzogiorno*

I first visited **Civita di Bagnoregio**, a town in the *Province of Viterbo* in central Italy, about 120 kilometers (75 mi) north of Rome, for the first time in 1991. The only access is a concrete pedestrian footbridge from the nearby town. I went there at the invitation of a couple who were responsible for much of the restoration of the town, Astra Zarina and Tony Heywood. Astra passed away several years ago and at the time of this writing Tony remained as one of its sixteen full-time residents.

Civita di Bagnoregio was founded by Etruscans more than 2,500 years ago. The street layout is of Etruscan origin, based on a cardo and *decumanus*, but the town saw its heyday during the Middle Ages and the Renaissance. I still remember an unforgettable evening cooking Cornish hens in Astra and Tony's enormous Etruscan chimney. There was more wine than hens that night. The next morning, the sight of the landscape through the window remains a treasured memory.

Civita's extraordinary appearance owes much to the tragic natural condition that has been gradually gnawing away at its core and foundations through erosion and landslides. Known as *"The Dying Town"* it remains an apparition in the mountains north of Rome. I have always been fascinated to document faded beauty in cities and in buildings. This obsession would inevitably draw me to **Venice** and **Sicily**.

When I first visited Venice, in 1974, my guidebook was Ezra Pound's Pisan Cantos and an extraordinary book of photographs by Vittorugo Contino called *Ezra Pound in Italy*. Pound had arrived in Venice in 1908 living over a bakery near the bridge of S. Vio. Years later, when I took my own students to that magical city, I returned with them to one of the treasures I had discovered then, the *Church of S. Maria dei Miracoli*, designed by Pietro Lombardo. Amid the shadows of the Venetian light the angels carved by his brother Tullio in 1481-89 were still there. Walking in Pound's footsteps, I stayed in Ruskin's *Pensione La Calcina*, and discovered the beauties of the *Cannaregio*.

No part of Italy has as many layers of history of different peoples as Sicily. From the Greeks who colonized *Siracusa* and *Selinunte*, to the Romans in *Agrigento*, to the Normans in Palermo. Its hybrid heritage is best encapsulated outside of Palermo in the *Palazzina Cinese*, a former royal residence of the House of Bourbon. It was designed in the *Chinoiserie* style in 1799 by the architect Giuseppe Venanzio Marvuglia on commission by King Ferdinand III of Sicily: A Chinese villa conceived by an Italian architect for a Spanish monarch! Such is the history of this extraordinary island.

My drawings are fugitive visions of this complex and multi-layered landscape. A homage to the one of the world's richest artistic traditions, and of my own transformative personal experience as an architect.

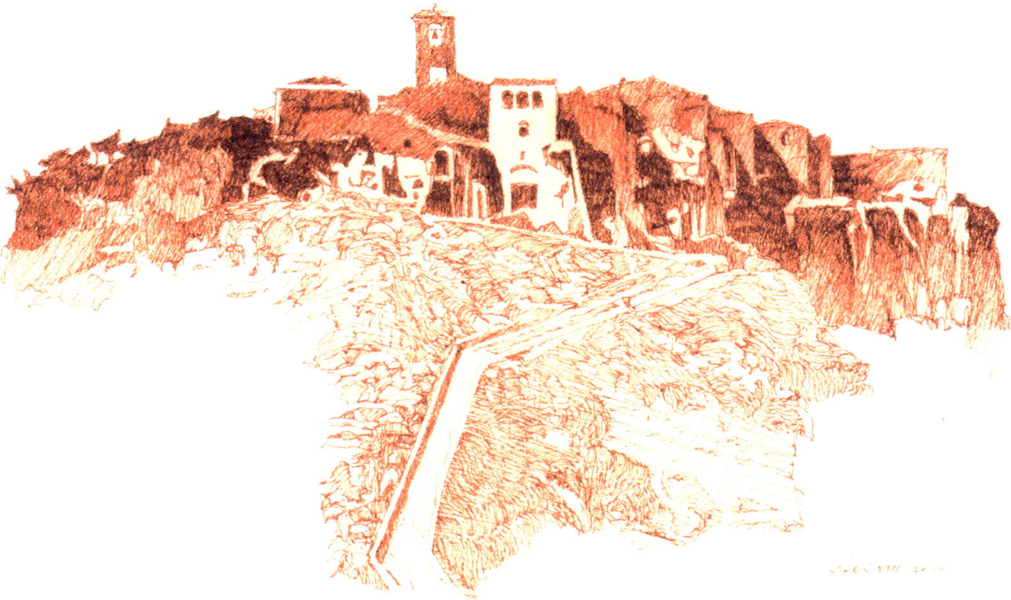

Civita di Bagnoregio, Viterbo
Felt-tip pen on watercolor paper

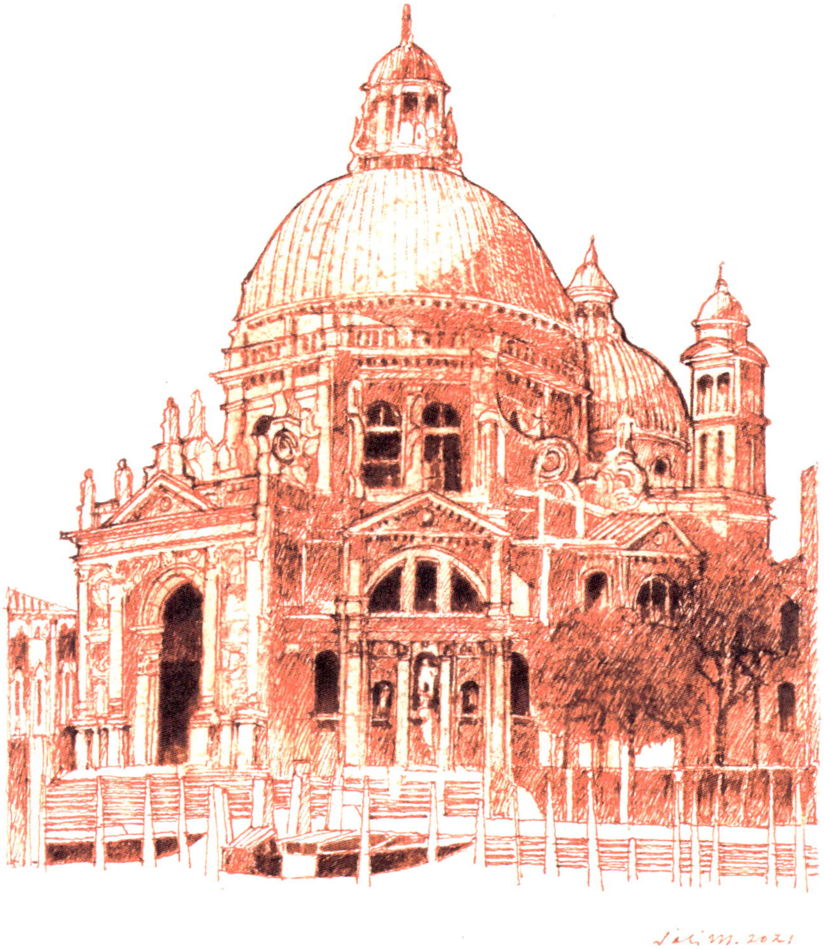

S Maria della Salute, Basilica, Baldassare Longhena, 1630, Venice
Felt-tip pen on watercolor paper

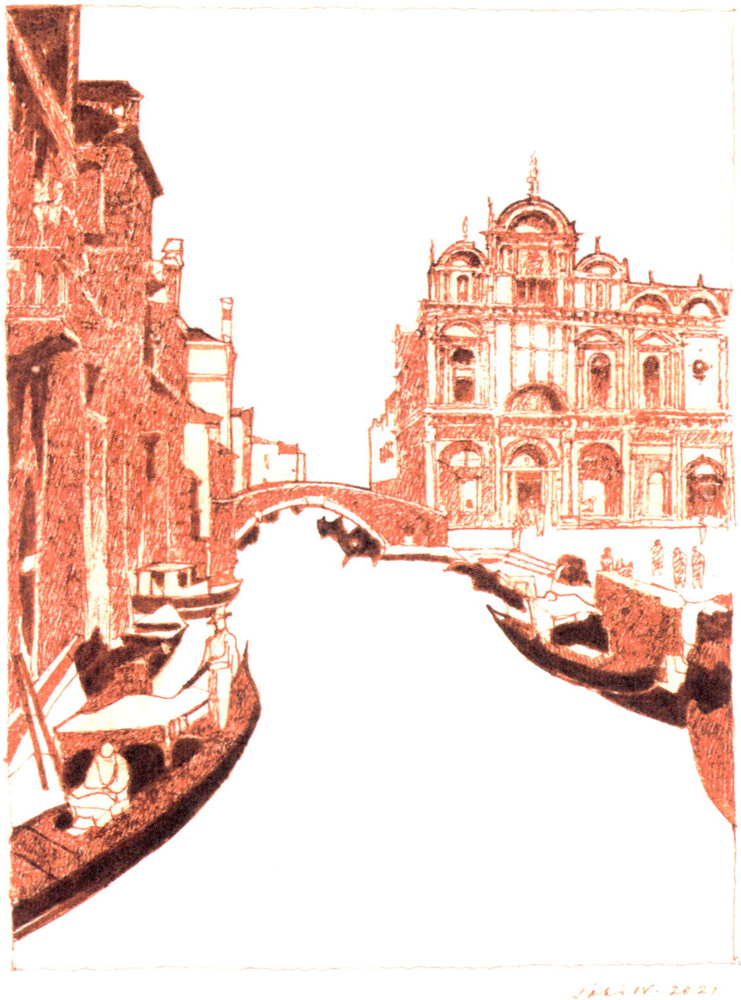

S. Giovanni e Paolo, Basilica, Venice
Felt-tip pen on watercolor paper

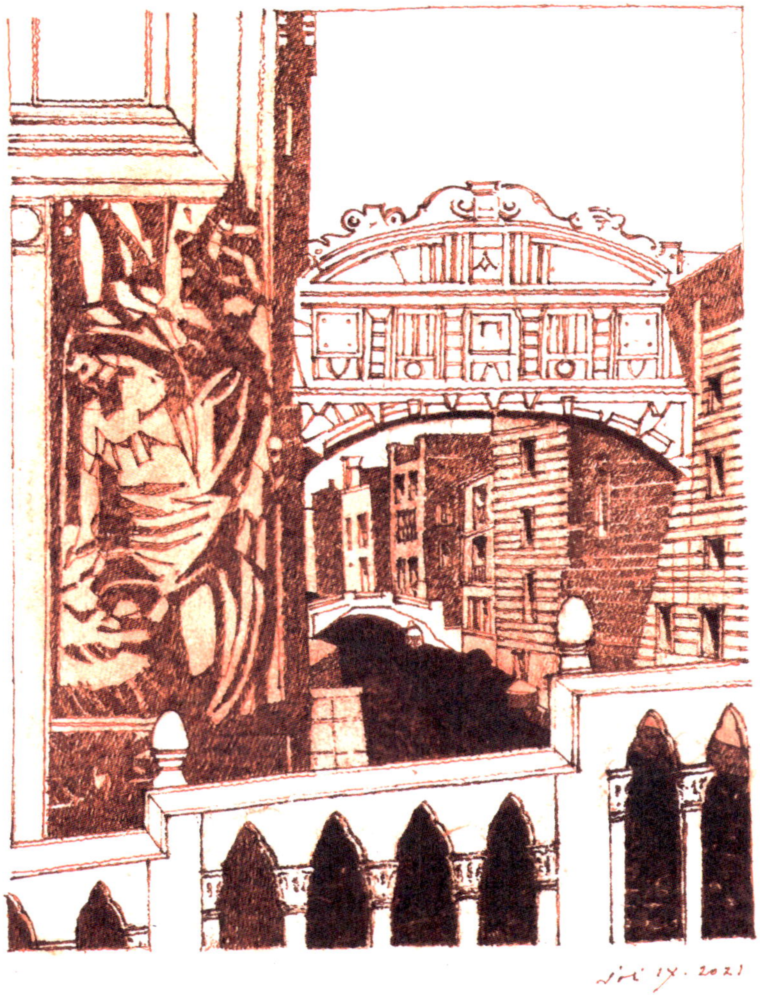

Bridge of Sighs, Antonio Contino 1600, Venice
Felt-tip pen on watercolor paper

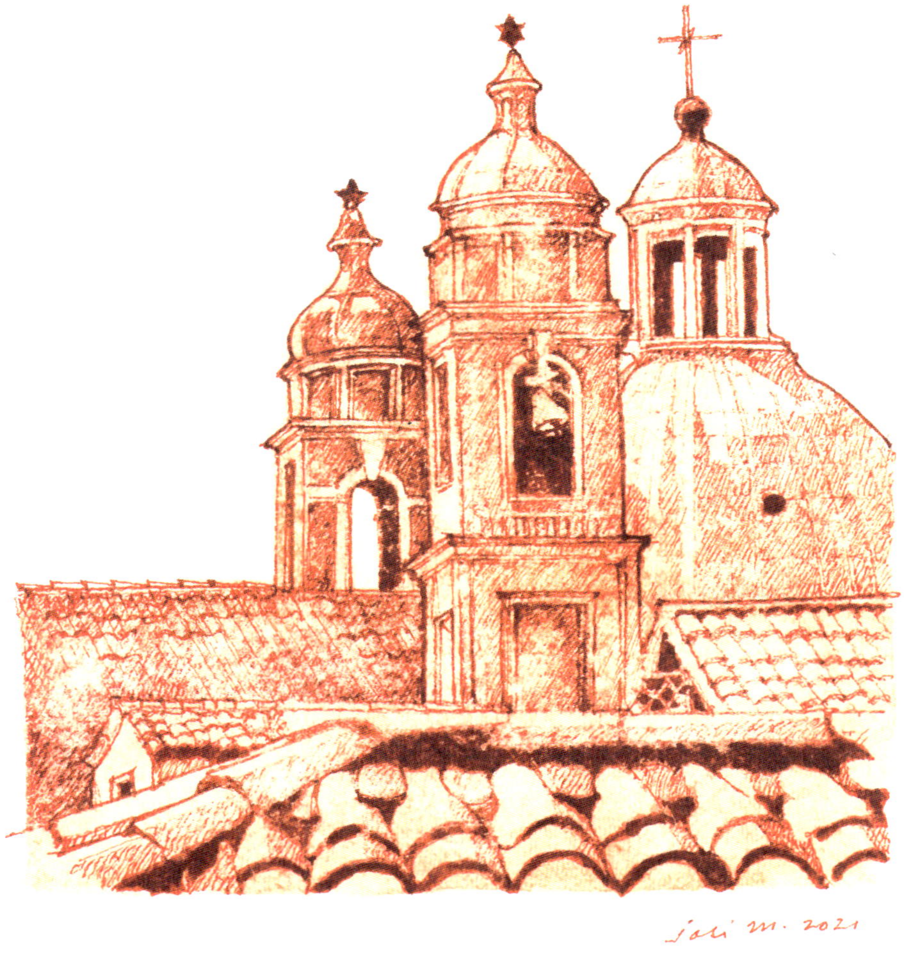

Domes, Venice
Felt-tip pen on watercolor paper

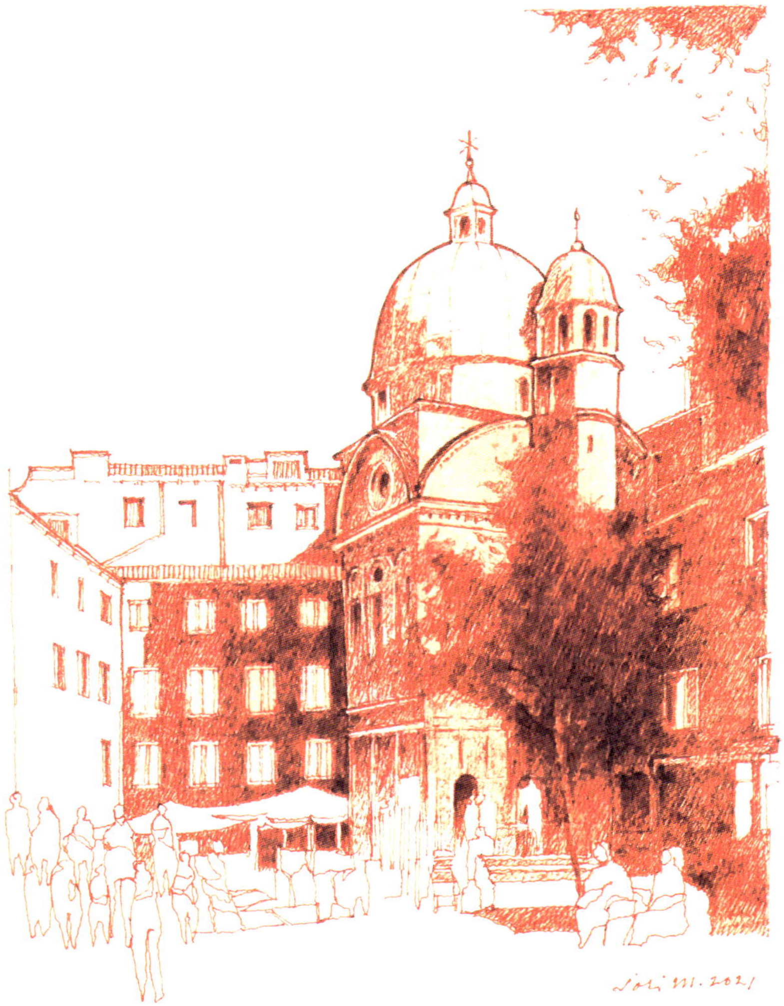

S. Maria dei Miracoli, Church, Pietro Lombardo, 1481-1489, Venice
Felt-tip pen on watercolor paper

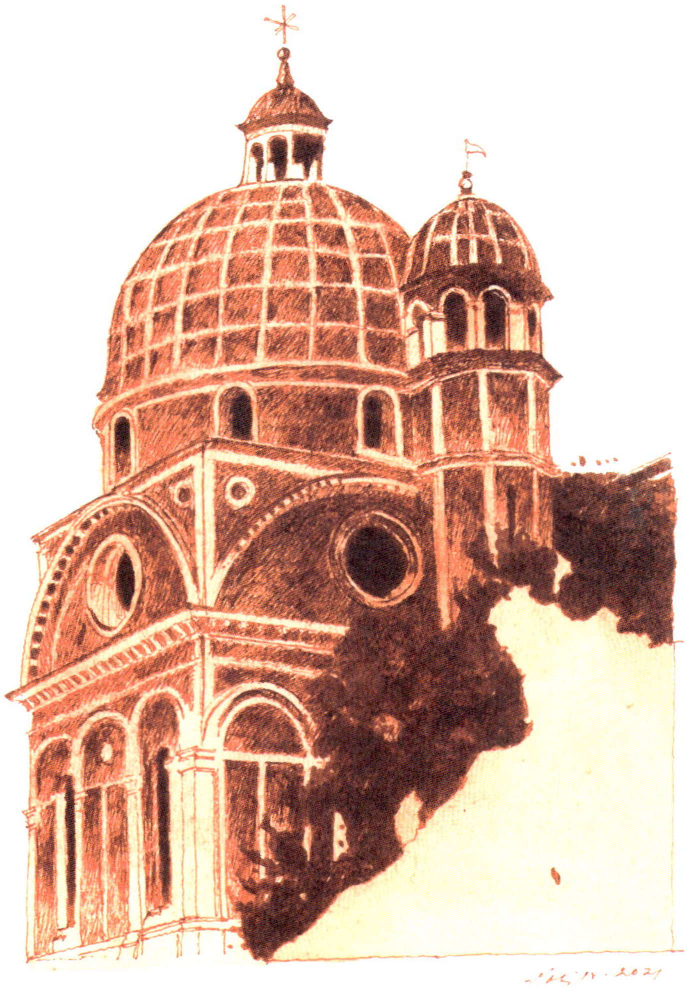

S. Maria dei Miracoli, Church, Pietro Lombardo, 1481-1489, Venice
Felt-tip pen on watercolor paper

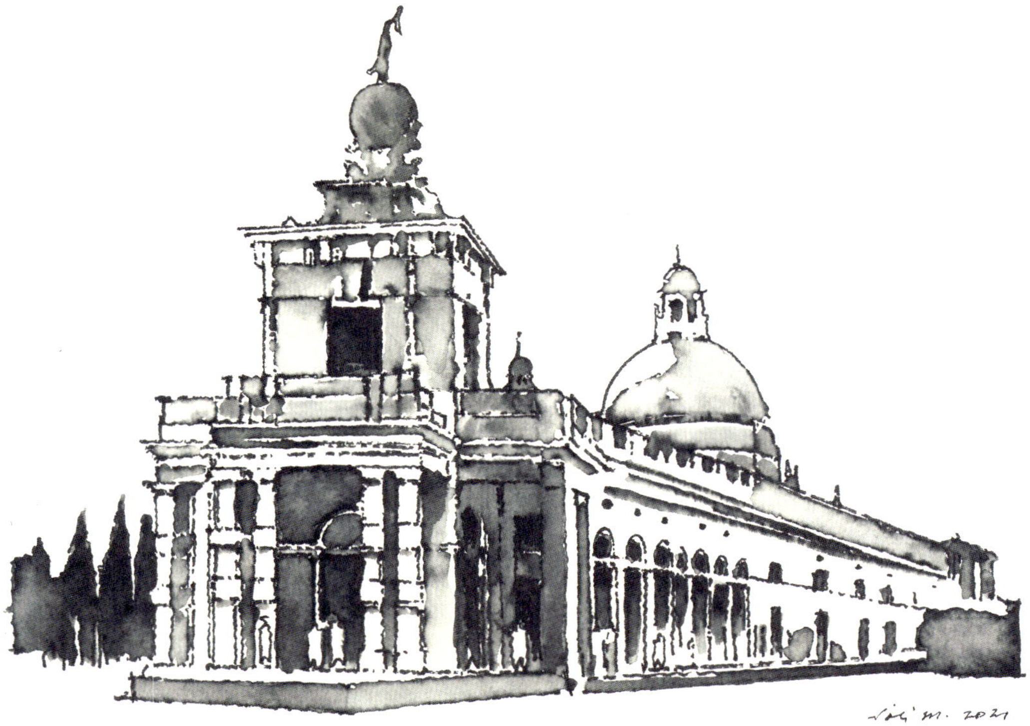

Dogana del Mare, Giusseppe Benoni, 1678-1682, Venice
Felt-tip pen on watercolor paper

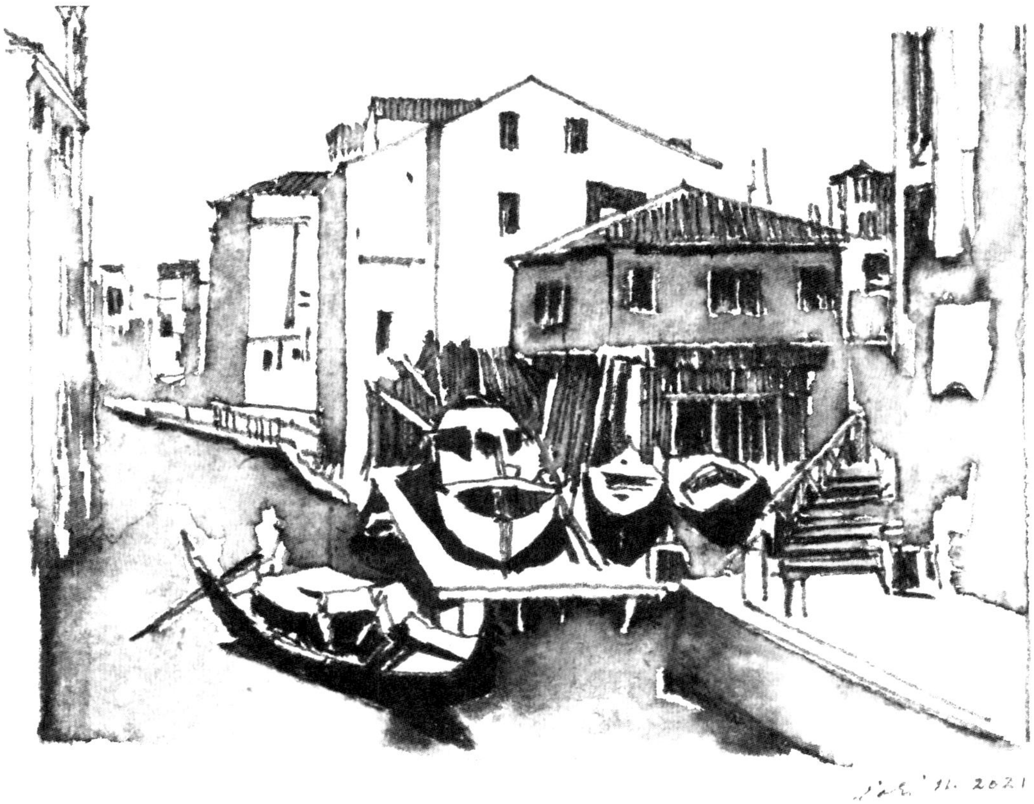

Canareggio, Venice
Felt-tip pen on watercolor paper

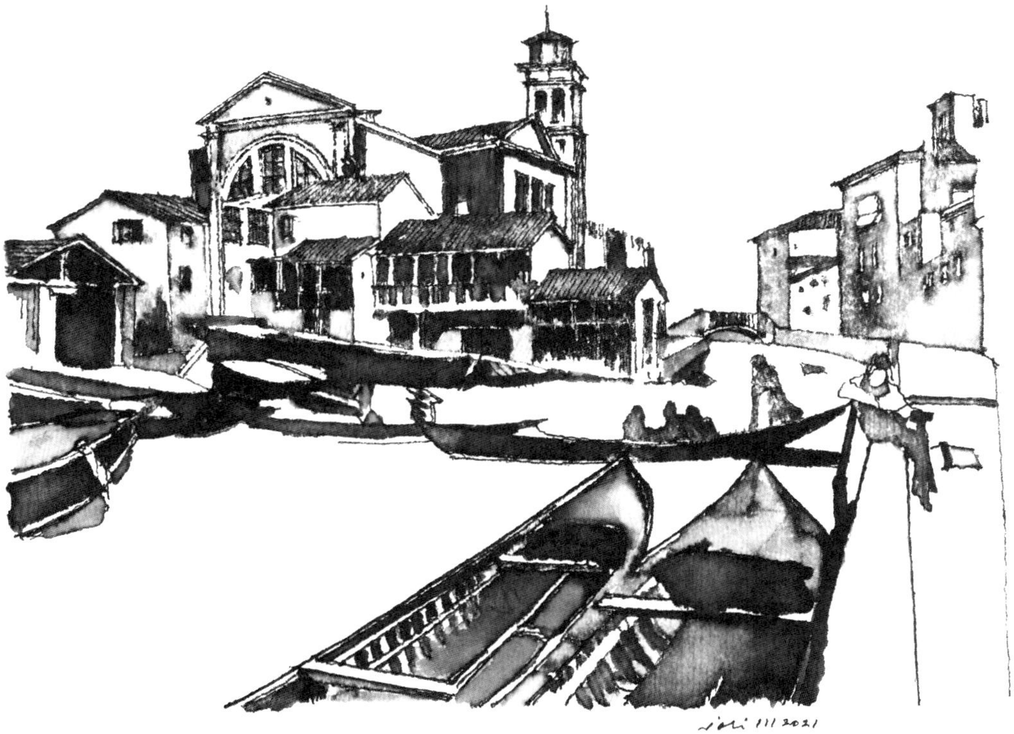

Canareggio, Venice
Felt-tip pen on watercolor paper

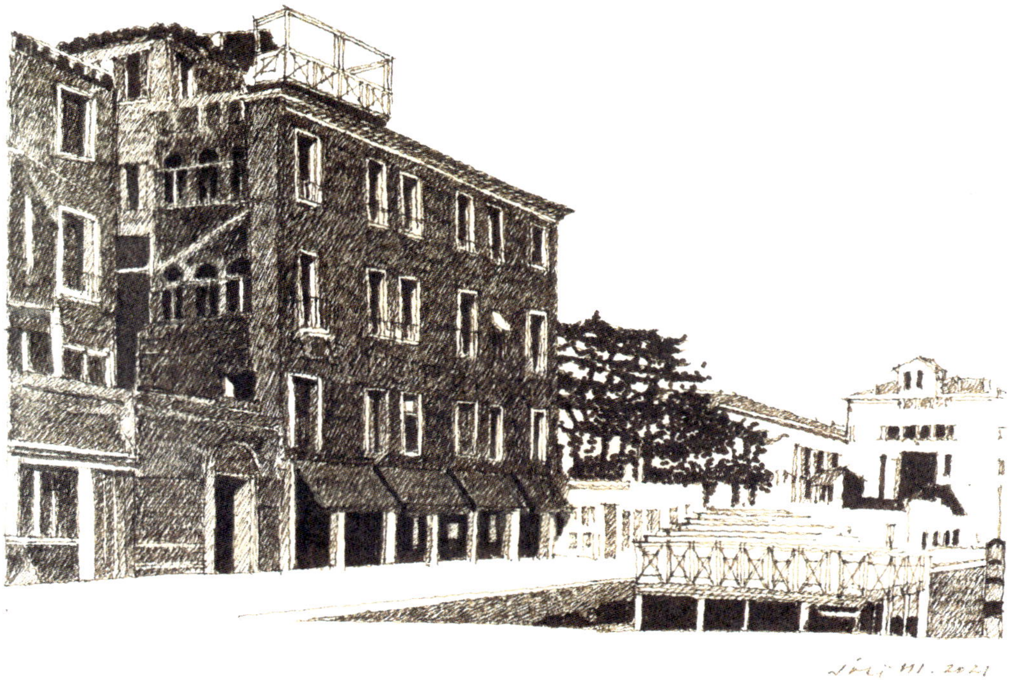

Pensione La Calcina, Venice
Felt-tip pen on watercolor paper

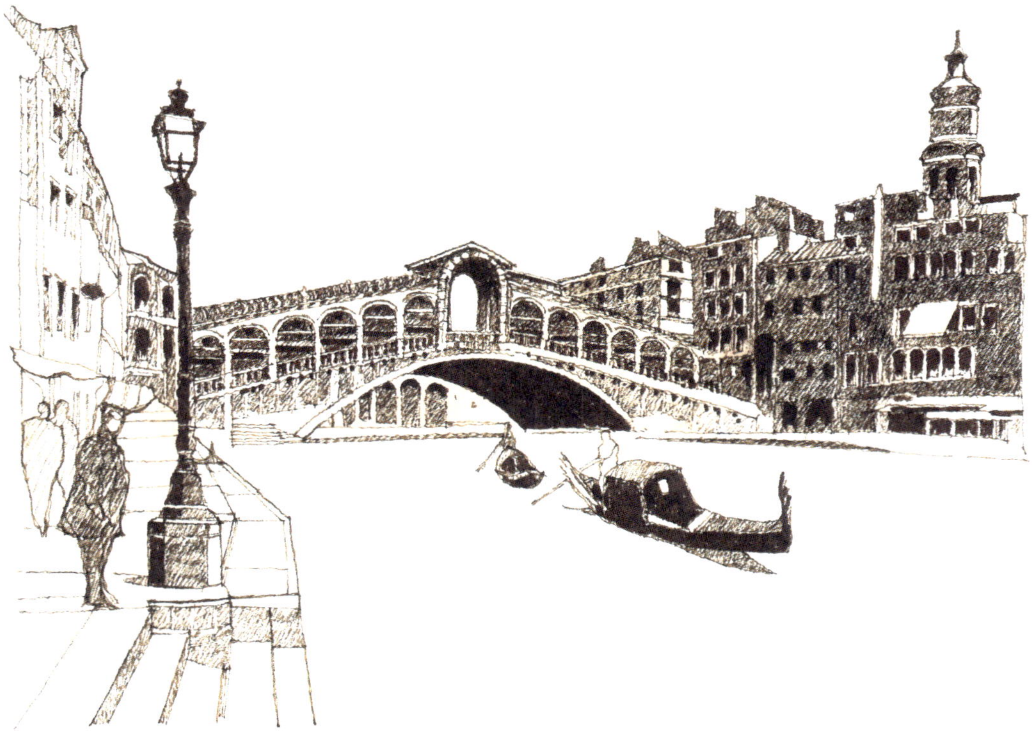

Ponte de Rialto, Antonio da ponte, 1588-1591, Venice
Felt-tip pen on watercolor paper

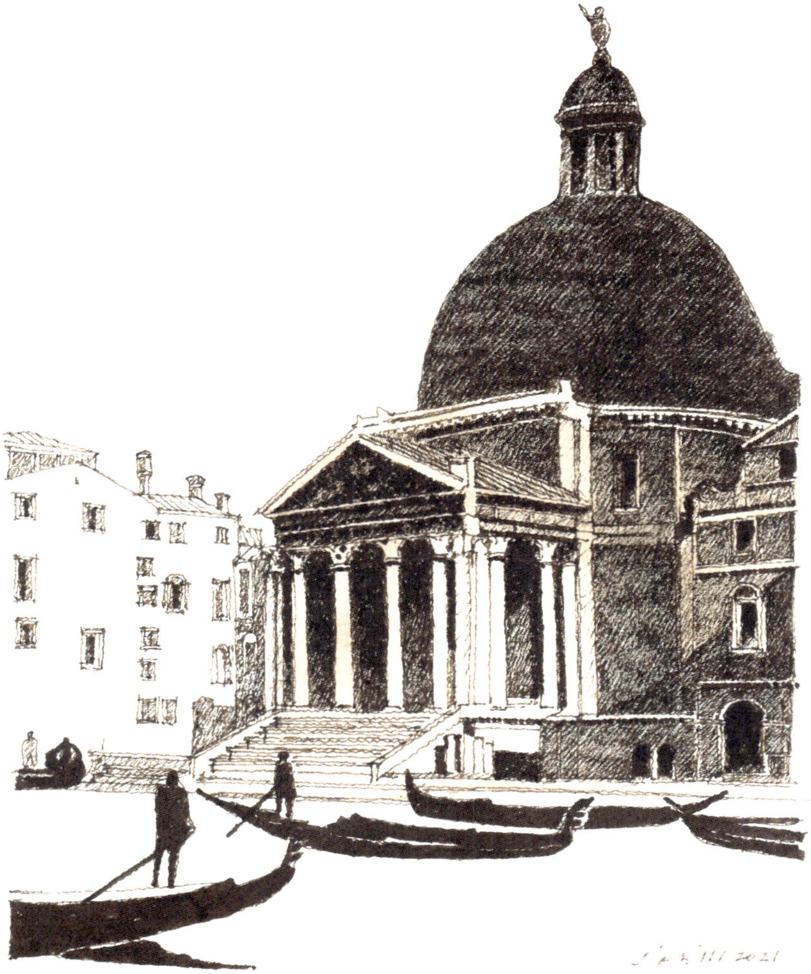

S. Simeone Piccolo, Giovanni Anotnio Scaltarotto, 1718-1738, Venice
Felt-tip pen on watercolor paper

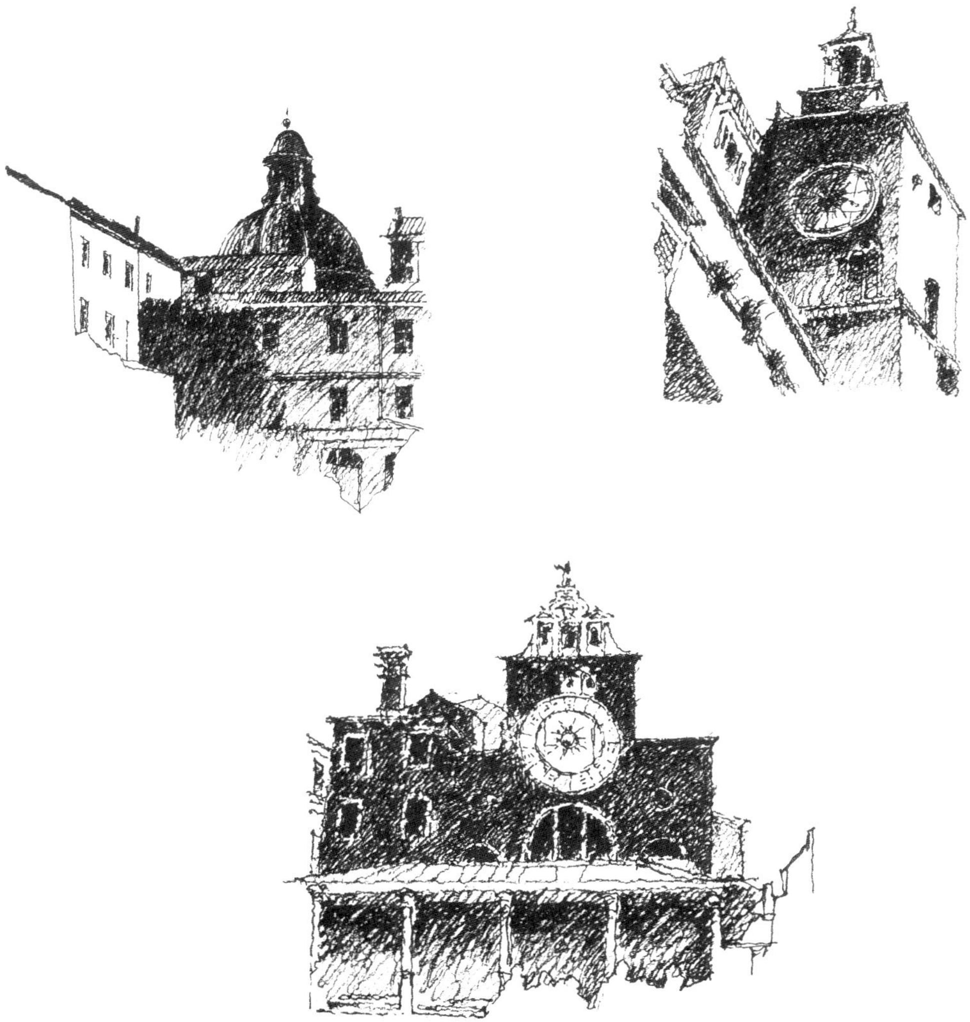

Campo della Pescaria, Venice
Felt-tip pen on watercolor paper

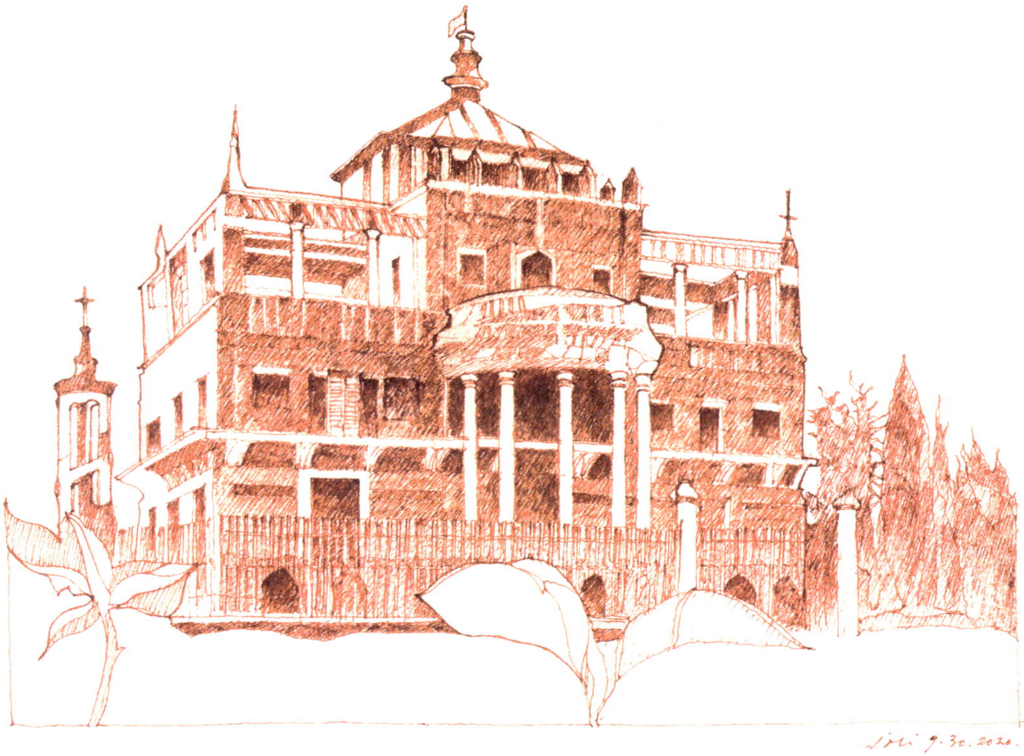

Palazzina Cinese, Venanzio Marvuglia, 1799, Palermo, Sicily
Felt-tip pen on watercolor paper

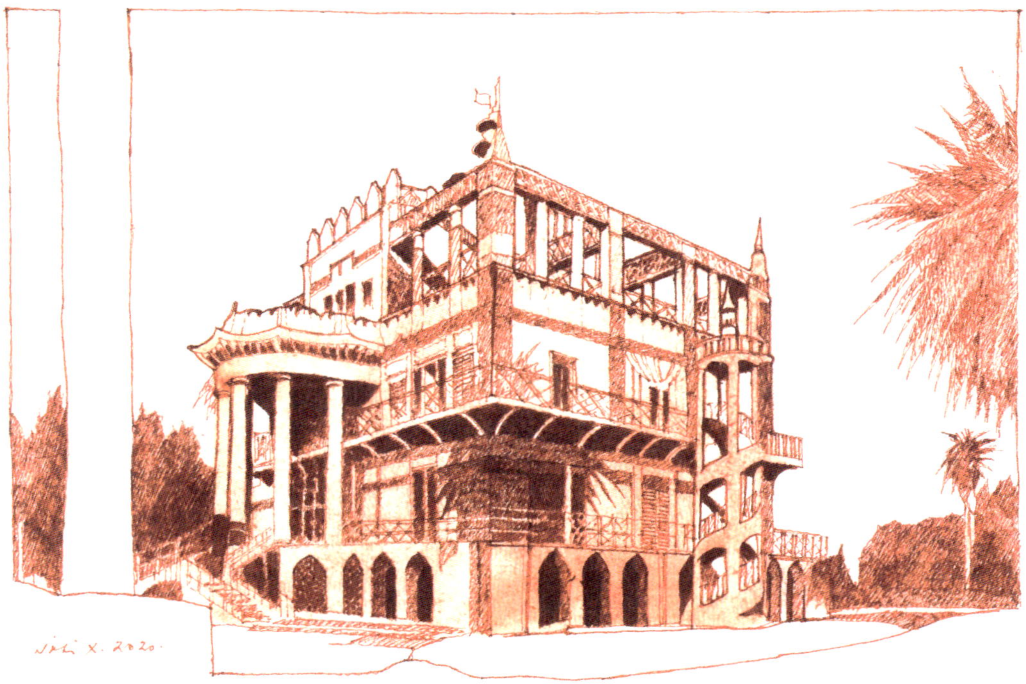

Palazzina Cinese, Venanzio Marvuglia, 1799, Palermo, Sicily

Felt-tip pen on watercolor paper

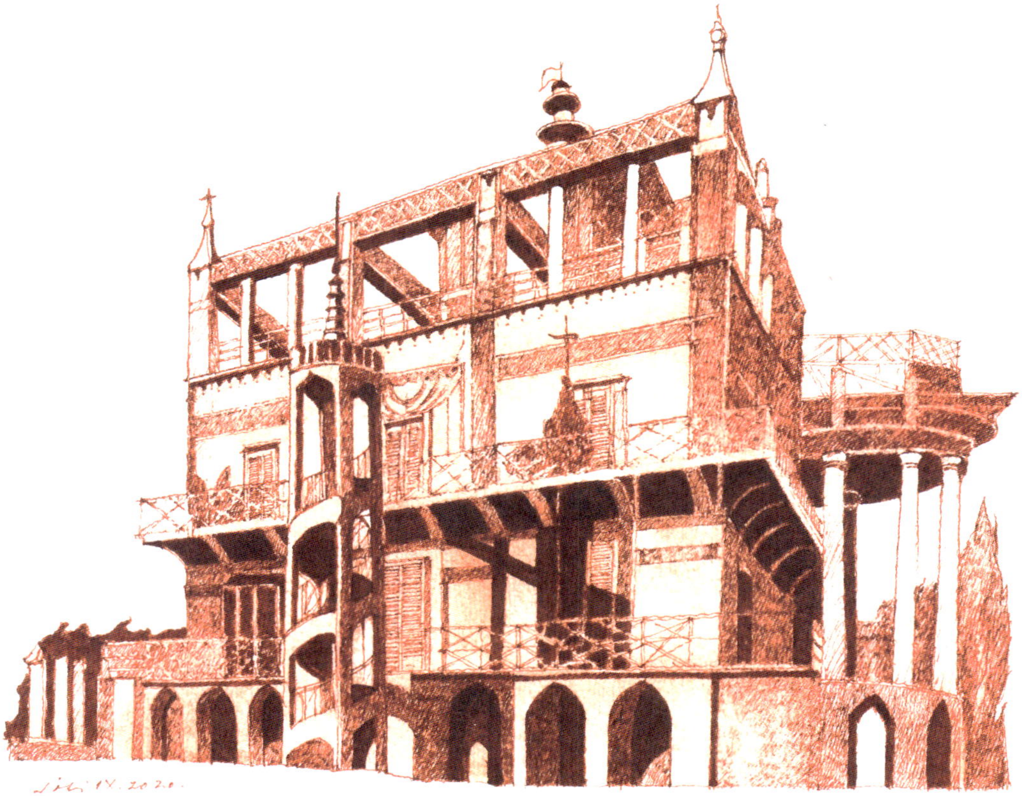

Palazzina Cinese, Venanzio Marvuglia, 1799, Palermo, Sicily
Felt-tip pen on watercolor paper

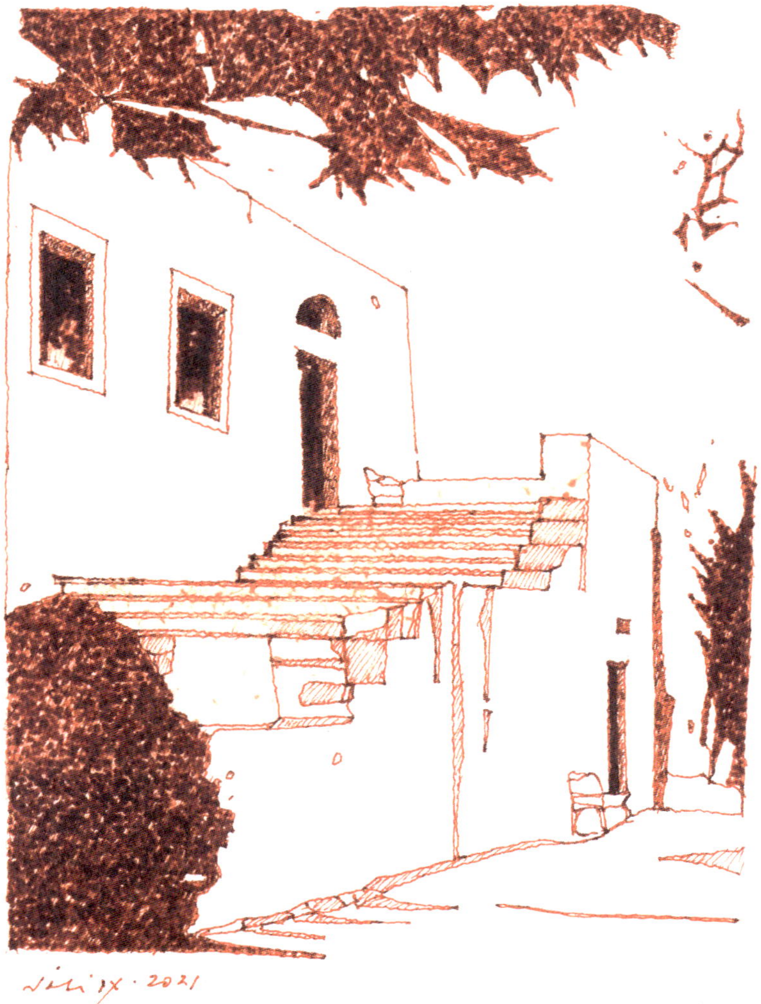

Pirandello House, Agrigento, Sicily
Felt-tip pen on watercolor paper

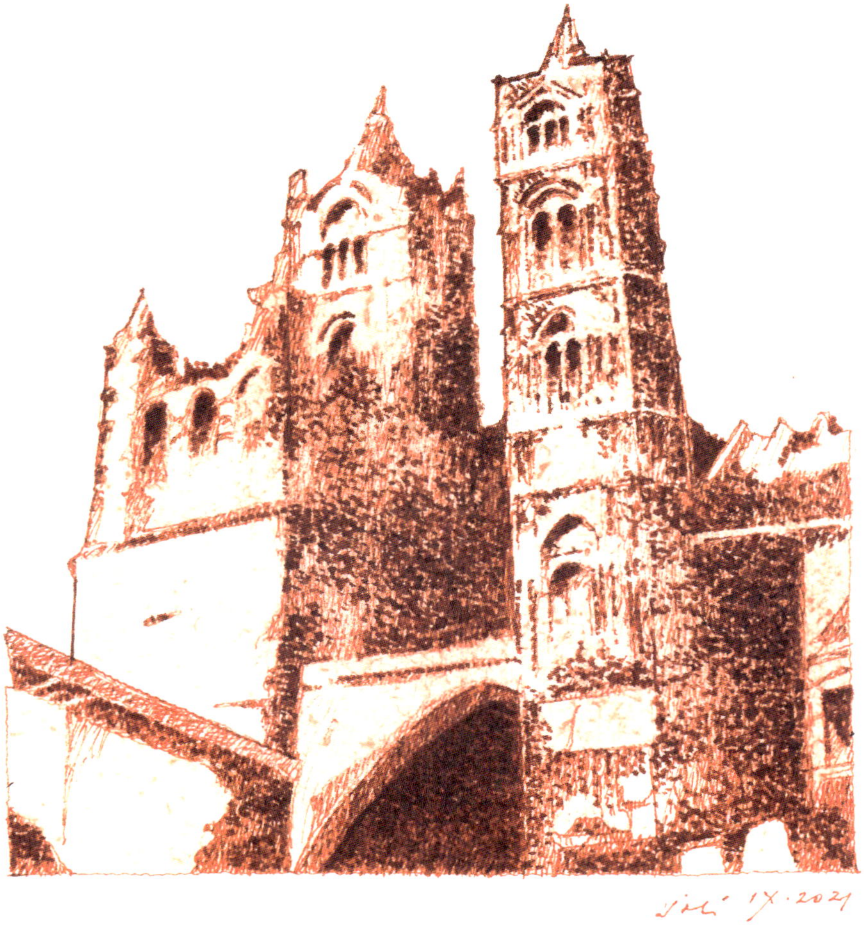

Cathedral, Ferdinando Fuga and Venanzio Meravuglia, 1185-1801, Palermo, Sicily
Felt-tip pen on watercolor paper

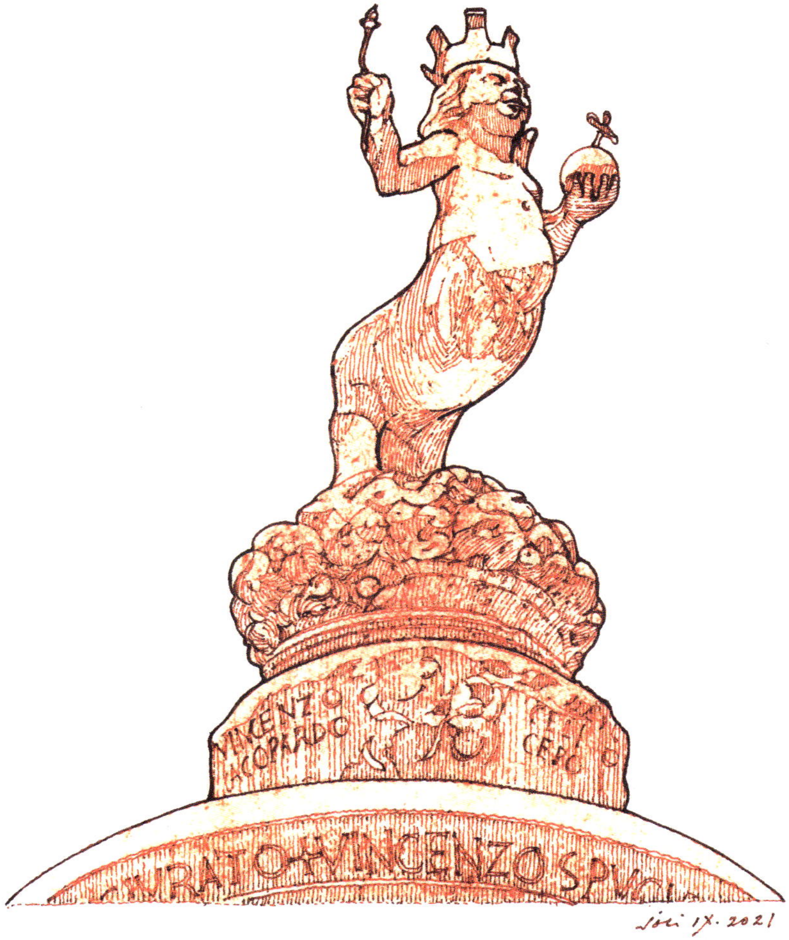

Minotaur in the Fontana del Duomo, Vicenzo Cacopardo, 1635, Taormina, Sicily
Felt-tip pen on watercolor paper

José Gelabert-Navia

José Gelabert-Navia is an architect, teacher and artist. Residing in Miami, Florida, Mr. Gelabert-Navia has over forty years of experience as an architect and is Regional Director for Latin America of Perkins+Will, Architects. Mr. Gelabert-Navia is currently a Professor of Architecture and formerly the Dean of the School of Architecture at the University of Miami.

A graduate of Cornell University, Mr. Gelabert-Navia's primary teaching focus is Architectural Design and History of Colonialism and Globalization in Architecture. He was the founder of the University of Miami School of Architecture's Rome Program and teaches a course in Italian Culture. He is the author of many publications, including *CUBA, Memories of Travel/Recuerdos de Viaje* (Oscar Riera Ojeda Publishers).

As the child of two architects, Mr. Gelabert-Navia was immersed in the world of architecture practically from birth.

Exile from Cuba shaped the path of Mr. Gelabert-Navia's career. He developed a global curiosity, moving from country to country, learning from each new location. These drawings are the records of those journeys.

This book is dedicated to my spouse, Amelia Rea Maguire. Thirty years. Eleven thousand days. No two the same.
And to Italy, where it all started.

Credits

Graphic design by Lucía Bauzá
Art direction by Oscar Riera Ojeda
Copy editing by Kit Maude & Michael W. Phillips Jr.

OSCAR RIERA OJEDA
PUBLISHERS

ISBN 978-1-946226-57-0
Published by Oscar Riera Ojeda Publishers Limited
Printed in China

Oscar Riera Ojeda Publishers Limited
Unit 1003-04, 10/F.,
Shanghai Industrial Investment Building,
48-62 Hennessy Road, Wanchai, Hong Kong

Production Offices
Suit 19, Shenyun Road,
Nanshan District, Shenzhen 518055, China

International Customer Service & Editorial Questions: +1-484-502-5400

www.oropublishers.com | www.oscarrieraojeda.com
oscar@oscarrieraojeda.com